OTB 2004

MW00723549

# PEOPLE AND INSECTS

Lynn M. Stone

The Rourke Book Company, Inc.
Vero Beach, Florida 32964

PHOTO CREDITS
© James H. Robinson: cover, p. 15, 19;
© J. H. "Pete" Carmichael: p. 4, 10, 12, 13, 20;
© Lynn M. Stone: title page, p. 7, p.8,
© James P. Rowan: p. 16

EDITORIAL SERVICES
Janice L. Smith for Penworthy Learning Systems

**Library of Congress Cataloging-in-Publication Data**

Stone, Lynn M.
People and insects / Lynn M. Stone.
p. cm. — (Six legged world)
ISBN 1-55916-311-9
1. Beneficial insects—Juvenile literature. 2. Insect pests—Juvenile literature.
[1. Beneficial insects. 2. Insect pests.] I. Title.

SF523.5 .S77 2000
595.7—dc21

00–036925

**Printed in the USA**

CONTENTS

## INSECTS AND PEOPLE

Over the centuries, we have learned many things about insects. For example, we know there are times when we don't want to live with them. We've also learned we can't live without them.

There are millions of **species**, or kinds, of insects. Naturally, some cause problems for us. Only about one kind in every 100, however, is harmful to us or our activities.

*An Atlas moth lands on the nose of a visitor to Florida's Butterfly World, where people walk among free-flying moths and butterflies.*

## INSECTS GOOD AND BAD

Insects really aren't "good" or "bad" in the sense that people are. Still, we often speak of insects as "bad" if they bite, sting, infect us with disease, or just, well, bug us! They're also bad, in our minds, if they attack our farm crops, pets, or property.

"Good" insects are those from which we get something good – a benefit. These helpful insects may delight us with their beauty. More likely, they produce honey or silk or help us with the growth of our fruits and vegetables.

*Many people enjoy insects because of their bright colors, strange shapes, and interesting habits.*

Certain insects, like bees, wasps, moths, and butterflies are called **pollinators**. Pollinators are extremely important in helping crops grow.

Pollinators move **pollen** around. Pollen is made up of tiny, sand-like grains found in flower blossoms. When an insect pollinator lands on a flower blossom to feed, pollen grains cling to it.

*A beekeeper checks his hive. Honeybees make more than enough honey for both themselves and the beekeeper.*

The pollinator moves from flower to flower. At each stop, it leaves – and picks up – pollen. The movement of pollen helps a plant **reproduce**, or make new plants. Honeybees, for example, pollinate fruit trees and other plants. Without bees, apple blossoms would never become ripe apples.

Insects are beneficial when they get rid of garbage by eating it. Some insects improve soil quality by tunneling and moving soil around. Insects are also an important part in the diet of thousands of other kinds of animals.

*This red and blue beetle will transfer yellow pollen grains from one flower to another.*

*This Costa Rican bullet ant's sting has been compared to being struck by a bullet!*

*The deep, piercing bites of deerflies are painful. Deerflies are pests to people, but wild animals eat them.*

## HARMFUL INSECTS

Some insects have become pests only with help from humans. Fire ants, for example, are not naturally found in the U.S. Fire ants came into the United States on wood shipped from Brazil. They had no natural enemies in the United States, so they spread throughout the South. Their digging has ruined farmland. Their bites have killed wildlife and tame animals alike.

*Fire ants swarm from their nest mound onto an intruder's rubber boot.*

Japanese beetles aren't native to North America either. In 1916 these shiny beetles were released in the United States by accident. They spread rapidly and gobbled flowers, fruits, leaves, and roots.

Most harmful insects are just pests. They damage books, clothes, wood, and food. Termites, cockroaches, clothes moths, silverfish, and carpenter ants invade people's houses.

*Harmful to human interests, Japanese beetles have spread in the United States after being released in 1916.*

Mosquitoes, ants, and bees ruin picnics. Weevils, earworms, and Colorado beetles destroy crops. And locusts can nibble a pasture clean. No-see-ums and a variety of flies and bees have nasty bites or stings.

Insect pests cause millions of dollars in damages each year. But some insects can be truly dangerous. They spread diseases such as typhus and **malaria**. Disease carriers include fleas, lice, and many fly species.

*Jaws of carpenter ants are not welcome in wood frame homes!*

## CONTROLLING INSECTS

Humans use many methods to control pests. **Quarantine** stations exist at borders between countries or states. They try to stop passage of products that might carry harmful insects.

Living controls are natural **predators**, or hunters, of insects. They are released to kill certain insects. Lady beetles were released in California to kill cushion scale insects. Until the lady beetles ate them, the native Australian scale insects had been killing orange trees.

*Ladybug beetle attacks aphids on a plant stem. Aphids eat plants, but ladybugs can help control aphid numbers.*

Chemical controls are widespread. They include products called **insecticides**. Insecticides kill large numbers of insects. But insecticides filter into ground and water. And they are often deadly to animals they weren't meant to kill.

In the 1960's, the insecticide DDT nearly wiped out ospreys, bald eagles, and brown pelicans in the lower 48 states.

Today many farmers try to control insects by changing their farming practices.

# GLOSSARY

**insecticide** (in SEK tuh sid) — a chemical used to kill insects

**malaria** (muh LAIR ee uh) — a serious, sometimes deadly tropical disease spread to people by certain kinds of mosquitoes

**pollen** (PAH len) — powderlike grains produced by some plants as part of their reproduction process

**pollinator** (PAH luh nay tur) — an animal whose flower-feeding activity shifts pollen from one plant to another, helping the plant reproduce

**predator** (PRED uh tur) — an animal that hunts and kills other animals for food

**quarantine** (KWOR un teen) — a means of separating or isolating something

**reproduce** (ree pruh DOOS) — to make others of the same kind

**species** (SPEE sheez) — within a group of closely related animals, such as beetles, one certain type (**Japanese** beetle)

## FURTHER READING

Find out more about people and insects and insects in general with these helpful books and information sites:


- Gibbons, Gail. *The Honey Makers*. Morrow Junior Books, 1997
- Jeunesse, Gallimard. *Bees*. Scholastic, 1997
- Rowan, James P. *Honeybees*. Rourke, 1993
- Stone, Lynn M. *Honey Farms*. Rourke, 1999

Wonderful World of Insects on-line at www.insect-world.com
Children's Butterfly Site at www.mesc.nbs.gov


## INDEX